Bibliographic information published by the German National Library:

The German National Library lists this publication in the National Bibliography; detailed bibliographic data are available on the Internet at http://dnb.dnb.de .

Imprint:

Print and binding: Books on Demand GmbH, Norderstedt Germany
ISBN: 9783668446748

This book at GRIN:

http://www.grin.com/en/e-book/358129/investigating-ambient-ai-techniques-suited-for-procedural-worlds

Mohamed Serry

Investigating Ambient AI Techniques Suited For Procedural Worlds

Ambient AI in Procedural Games

GRIN Publishing

Investigating Ambient AI Techniques Suited for Procedural Worlds

Honors Thesis for B.Sc.(Hons) In Games Programming

UNIT: GAMES 620 (HONS)
STUDENT: Mohamed Serry

TABLE OF CONTENTS

Acknowledgements

Thejaswi Padmanabha: firstly, I'd like to thank Mr. Thejaswi for his excellent supervision and assistance throughout my honors period at SAE. He consistently provided thoughtful and inspiring guidance that helped me scope my research in the correct direction, he consistently believed in my efforts despite losing sight of my direction a number of times. Furthermore, He often gave me advice beyond academic matters when it seemed to me that the difficulties I was facing were of a personal nature. I would arguably say without his consistent faith in me this research would not have been a successful endeavor.

Cloderic Mars: secondly, I'd like to thank Mr. Cloderic Mars for his initial assistance during the preproduction phase of my research. Despite his limited time, he was able to answer my questions and proving industry insight into my work.

List of Tables

List of Figures

Abbreviations

AAI	Ambient AI
NPCS	Non-Playable Characters
PCG	Procedural Content Generation
IGD	Independent Game Developer
PE	Procedural Environment
RPGS	Role-Playing Games
RTCG	Run-Time Content Generation
GAIC	Game AI Conference
AL	Artificial Life
HFSM	Hierarchical Finite State Machine
AALS	Ambient Animation Loops
CiF	Comme il Faut
NLP	Nature Language Processing
LoD	Level of Detail
CCS	Context Correct Spawning
PGC	Procedurally Generated Content
PG	Procedural Games
PC	Procedural Context
FPS	Frames Per Second
ms	Frames per millisecond

Investigating Ambient AI Techniques Suited for Procedural Worlds

Abstract

Procedural Content Generation (PCG) is a trending technique in content authoring for Independent Game Developers (IGD), recent success in games such as *No Man's Sky* has led developers to investigate this newly developed technology.

Now more than ever, Game AI has to adapt to a changing environment and delivering a unique gaming experience on Run-Time. A selection of Ambient AI (AAI) Techniques is selected for identifying the most suited technique for Procedural Games (PGS). In the selection process, a prototype of each AAI technique that illustrates its core concepts is simulated and tested in a semi – generated environment that resembles certain elements in a PE. Furthermore, each technique will be subject to testing sessions in order to evaluate the effectiveness of the approach of each technique based upon comparisons conducted between the simulations.

Research Statement

This research aims to identify effective AAI Techniques suited for simulating Background Ambience in PE.

Research Methodology

Research is conducted through the usage of both qualitative and quantitative methods.

- **Quantitative:** a simulation of each AAI will be conducted in order to derive data that can be utilized for quantifying its effectiveness in a PE.

- **Qualitative:** after the quantification of each AAI technique, it shall be further qualified by arguments that hypothesize its effectiveness by a thorough comparison of a selection of AAI Technique.

Research Scope

The emphasis of this research is to identify the most effective AAI technique for background Ambience in PE. The metrics for the solution shall constructed based on hypothetical scenarios that are inspired by ambient environments present in RPGS, where an abundance of NPCS are present in a simulated setting.

Research Limitations

The research data is collected from simulations conducted in the Unity3D Engine. As a result, the simulations will be confined to the engine's design which may introduce bias as some simulations could be more compatible than others. Moreover, the research is not conducted in the context of a commercial game. As a consequence, the data gathered could be less comprehensive as each AAI Technique is only simulated through a test scenario. However, specific elements of procedural environments have been considered to allow the simulation results to be in some degree closer to a full-fledged procedural scenario. For instance, the application shall make use of RTCG to resemble the setting of PE.

AI in Games: Introduction

Preceding the research, a thorough definition of the research field is required. The first research enquiry is what is Game AI? Game AI is generally considered as a tool for immersion, it focuses on creating an illusion of intelligence and creating a particular experience for the user. As a consequence, the techniques used in game AI and generally any other field employed in games is to bring about the desired experience for the user (Kevin Dill, 2014).

Defining Ambient AI in Games

To begin with, according to David Rez Graham in Game AI Pro, Ambient AI is an important component of any game that includes NPCS as it adds to the believability and immersion of the game (Graham, Breathing Life Into Your Background Characters, 2015). Moreover, according to Jeet Shroff in Game AI Pro 2, building believable NPCS is a core goal of game developers overall. Additionally, believable NPCS compel the user's interaction. Therefore, further grounding the user in the game world and selling the game's story (Mike Lewis, 2015). Likewise, according to Cloderic Mars in Game AI Pro 2, designing Ambient NPCS that breathe life into a virtual environment is a method for improving the feeling of presence for the user (Cloderic Mars, 2015).In Conclusion, Ambient AI is a collection of AI Techniques used for simulating believability in a virtual environment as a whole.

The Role of Ambient AI in Games

In this section, a case study of two games is conducted in order to deduce the role of Ambient AI in games. Firstly, based on thorough investigation an upcoming game named "Kingdom Come: Deliverance" developed by Warhorse Studios is selected. Secondly, a released game named "Far Cry 4" developed by Ubisoft Montreal.

According to Julien Varnier, Lead AI Engineer on Far Cry 4 in his GAIC 2014 presentation, the AI is divided into 4 main categories: Enemy AI, Friendly AI, Civilian AI and Wild Life (Varnier, 2014). In FarCry4 the civilian AI and the wild life are the applications of Ambient AI. In Continuation, the civilian AI is for populating FarCry 4's domestic areas and potentially providing interesting situations during their interactions, further enhancing the player's experience (Varnier, 2014). Moreover, the wild life is the major contributor to the overall ambience as the user frequently interacts with them as part of the gameplay and they consistently interact with other NPCS.

However, they also play as a supporting role as the civilian NPCS in breathing life into the environment as FarCry4 is set in a forest environment (Varnier, 2014). Nonetheless, Enemy AI is the major category as the player constantly interacts in conjunction with FarCry 4's Genre. In Contrast, according to Thomas Plch of Warhorse Studios in his presentation on the AI Engine behind Kingdom Come: Deliverance during at GAIC 2014, Ambient AI was the central focus in the game as it aims to achieve realism (PLch, 2014), opposite to player entertainment as FarCry 4. Moreover, the world to be populated is not open world as in the case of FarCry 4, this leads to more available resources for the simulation.

In Conclusion, Ambient AI portrays complementary roles in games often in support of other Gameplay Ingredients. Furthermore, the goal of the game clearly dictates the level of ambience required.

Ambient AI Techniques: An Overview

To further understand the role Ambient AI in games, a selection of Ambient AI techniques is used for preliminary analysis. These techniques are: Artificial Life (Schwab, 2008), Scheduling (Kevin Dill, 2014), Comme il Faut (CiF) (Kevin Dill, 2014), Smart Zones (Mike Lewis, 2015) and Ambient Animation Loops. The criteria for selecting these techniques in contrast to others is that each of these techniques is an Ambient AI Technique and can be applied or relates to the research criteria.

AL

According to Brian Schwab, Lead AI Engineer at Magic Leap Inc. in his book AI Engine Programming, AL is the attempt to simulate biological phenomena in a virtual environment, it is also the name attributed to a collection of disciplined that are employed together to formulate such a simulation (Schwab, 2008).

Furthermore, an advantage of this technique is that it results in emergent behaviors in a semi-consistent fashion. Moreover, it enforces reusability of behaviors since it requires governing rules to dictate the how the simulation will be conducted. Therefore, the gameplay must be divided into building blocks until it can be used as a rule (Schwab, 2008).

However, emerging behaviors from this technique are un-predictable as they rely on biological mechanisms that may take any shape or form during development, resulting in a critical issue should the Agents in the simulation start developing unwanted behaviors. Additionally, it can be difficult to tune or adjust parameters in the game as it is not clearly understood how they influence one another in the presence of poorly understood biological mechanisms in the simulation (Schwab, 2008).

Scheduling

According to David "Rez" Graham" in his previously mentioned article, Scheduling is composed of a HFSM that is operated by defining time blocks where each background NPC is to execute an action at the specified time (Graham, Breathing Life Into Your Background Characters, 2015).

Furthermore, each NPC is given a schedule, it can be thought of as a black box where the input is the current time and the output is an action the NPC can take, unlike AL scheduling is not for simulating the ambience of life but rather to give the *illusion* of its presence. In reality, the actions may be as simple as navigating to a certain point and playing an animation (Graham, Breathing Life Into Your Background Characters, 2015).

CiF

According to Michael Mateas and Josh McCoy in their Game AI Pro article titled "*An Architecture for Character-Rich Social Simulation*", CiF is an AI Architecture that aims to focus on simulating social activity between NPCS. it was employed in the development of Proom Week, a social simulation game where the player interacts with a group of high school students.

CiF aims to represent social interactions between NPCS through a collection of Dialogue exchanges between the agents. It utilizes NLP for dynamically generating dialogue texts for each character. Furthermore, in CiF social interactions between NPCS are based on Erving Goffman's work in dramaturgical analysis of social interaction in sociology (E.Goffman, 1959) which in summary is simply viewing social interactions in everyday life as a dramatic performance. Therefore, Interactions are based on what element of the game's social state are they designed to express or change. However, CiF is mainly designed around the need of a dialogue system, which may not be mandatory for most games (Mike Lewis, 2015).

Smart Zones

According to Cloderic Mars in his Game AI Pro 2 article titled "*Smart Zones to create the Ambience of Life*" smart zones are a newly developed concept that aims to design creditable, consistent and interactive ambient life (Mike Lewis, 2015). Smart zones are composed of "*Living Scenes*". A living scene is a set of NPCS interacting with and possibly with players. Furthermore, living scenes are designed to fit in the context of their virtual environment (Mike Lewis, 2015).

Each NPC participating in a scene fulfills a *role*, each role is defined by a set of behaviors that will be executed by the NPCS to achieve the scene. The behaviors assigned to each role are orchestrated in order to achieve the collective ambience required of the living scene. Furthermore, for executing a living scene, it is defined through a smart zone. Smart zones are a standalone module present in the virtual environment that is responsible for managing the lifetime of one living scene. Moreover, the aim of smart zones is to split the design of AAI into two individual layers; individual NPC behaviors and collective living scene behaviors (Cloderic Mars, 2015).

AALS

Perhaps the most classical AAI Technique, according to David "Rez" Graham it is composed of NPCS looping animations over and over. Each NPC loops a certain animation or action repetitively such as walking around a specific region. In combination it gives the illusion that each NPC is doing something different (Graham, Breathing Life Into Your Background Characters, 2015)t.

PE and AI: An Overview

In this section, an investigation of possible challenges for developing AI in a PE is conducted, based on investigation the case study selected is a recently developed procedural game named "*A Valley Without Wind*", it was developed by Arcen Games.

According to Chris Park lead developer of the game in his live interview with Alex J. Champandard of AIGameDev.com (Park, 2012) the entire game is procedurally generated. It is composed of a world map that's in 2D top-down view, the world map is dissected into sections called "regions" which are procedurally generated one part at a time as the player progresses through the game. Furthermore, each part of the game itself is constructed from layers of procedures that are combined together to increase the LoD in the game and the richness of the generated content.

On a high level, the main candidate for issues concerned with AI is the random nature of the generated environments. Furthermore, this high level issue is broken down into segments. To begin with, the first aspect of this issue is Context Correct Spawning (CCS). According to Chris Park, during the development of the game they faced difficulty correctly spawning entities in the right context. For example, an issue was reported during play testing that a T-Rex was spawned in a narrow corridor and had difficulty navigating (Park, 2012).

As Chris Park continues, it seems to be a problem with how procedural generation works rather than an AI Problem. The issue was addressed by utilizing a set of filters to determine if a spawn location is feasible for the type of entity being spawned (Park, 2012). In Conclusion, AI in a PE is required to adapt to a new context as the environment is generated. Furthermore, easily integrating diversity in contrast to scripted behaviors or pre-defined contexts is a key factor.

AI and PE: Further Case Studies

Building on the previous case study, potential criteria for AI in a PE could be derived. Firstly, it was mentioned that AI would be required adapt to fit a new context continuously. Moreover, a secondary case study on the usage of PCG in RPGS is surveyed for based on a qualitative investigation of literature on the topic of PCG, the selected case studies reflect the usage of PCG in RPGS which is the target theme for potential simulations based on this investigation.

To begin with, according to Nathaniel Buck in his bachelor's thesis titled "*Procedural Content Generation in Strategy/Role-Playing Games*" PCG is used to completely generate levels on run-time, due to that, the theme and story of the game cannot be pre-defined or yield an expected outcome, instead only the major events are linearly defined (Buck, 2013).

Moreover, according to Tam Adams in his article titled "*Simulation Principles from Dwarf Fortress*". Dwarf Fortress is an RPG built on simulation, the game world is procedurally generated from scratch over multiple steps. Tam states that given adequate CPU and memory resources, a deep and rich simulation can be achieved through its generated content. Therefore, it can be assumed that PGC is potentially consuming in terms of computational resources. Indubitably, this would rely on the type of simulation being developed. However, PG may allow for less computational resources for Gameplay Ingredients as they are expected to be generated to some extent rather than authored. On these grounds, secondary elements of AI such as AAI may be required to place a considerable emphasis on adaptability and performance. Lastly, according to David "Rez" Graham in his previously mentioned article, this hypothesis is further emphasized by the following: in order to decide on a strategy for AAI the most relevant criteria to be considered are: high performance, scalability, ease of implementation (Graham, 2015).

In conclusion, AAI Techniques would require to be efficient, scale able and easy to implement as well as being adaptable to a new context on run-time. Based on this hypothesis, a further analysis of the selected AAI Techniques is due.

AAI Techniques: Further Analysis

this section each of the previously mentioned techniques are reviewed based on the proposed hypothesis.

To begin with, according to Brian Schwab in his book "*AI Engine Programming*" AL operates on preset rules to simulate biological processes akin to life in a virtual environment. Furthermore, the rules that govern the simulation must not change as this would conflict with the predefined goal of AL simulations and would not depict accuracy in the results. Furthermore, AL techniques by default require a fixed context which directly conflicts with the nature of a PE. Therefore, AL may no longer be a viable candidate for experimentation as it is valid according to the concluded hypothesis on criteria of AAI Techniques in a PE.

On the other hand, according to David Rez Graham in his previously mentioned article. a full-fledged AAI system was developed utilizing a scheduling approach and later employed in a commercial title "*The Sims: Medieval*" and the solution was successful in achieving the following criteria: ease of implementation as it gave game designers less constraints to handle when designing background ambience as well as high efficiency as they were able to simulate a large number of NPCS. More Importantly, scheduling can be adapted for player integration with sufficient changes (Graham, 2015).

In Continuation, CiF was developed to be functional in a dynamic context as well, according to Micheal Mateas and Josh McCoy the dialogue patterns were developed in a generic manner that did not depend on any specific NPC. Rather they were developed as social settings that can occur between any two NPCS engaged in dialogue. However, the dialogue patterns themselves were pre-defined. Moreover, CiF like the Sims series is developed for use in simulation settings similar to AL such that the target ambience is not background ambience, whereas RPGS are rather directed experiences in which AAI is often employed for supportive roles as previously mentioned. This possibly places CiF out of the scope of the potential AAI Techniques that can effectively be used in a RPG setting.

Similarly, according to Cloderic Mars in his previously mentioned article, smart zones orchestrate behaviors between NPCS based on a set of pre-defined "roles" assigned to each NPC upon entering a smart zone. However, unlike CiF the roles are context-independent. For example, a smart zone may require the role "waiter" to be casted but it does not define who it is, it simply queries the world in search for viable NPCs and if none is found then the smart zone does not begin the living scene execution (Cloderic Mars, 2015). This is a possible indication that smart zones may be adaptable on run time.

Lastly, as previously mentioned by David Rez Graham, AALS is simply repeated behavior cycles where ambience is achieved through quantity (Graham, 2015). Furthermore, due to its simplicity it is quite efficient as it is merely a repeated animation. However, looped animations often heavily depend on the context in which they appear, Nonetheless, with the usage of CCS it may be a considerable candidate for investigation in a PE.

In conclusion, after further analysis of the proposed techniques, CiF and AL are no longer regarded as valid candidates due to the lack of qualitative evidence that they may be utilized well in a PE for simulating background ambience in light of the recently established hypothesis.

AAI Techniques: Selection Criteria

The proposed criteria for measuring the effectiveness of each AAI Technique are the following: High Performance, scalability, ease of implementation and run-time adaptability. However, **ease of implementation** is quite subjective and potentially dependent on which platform is the technique applied. Therefore, in order to avoid possible bias, it shall not be considered. Lastly, further refinements are required for classing the established criteria based on the defined research methodology.

To begin with, our first and potentially most important criterion is **high performance.** In order to measure the performance a bench mark must be defined. According to an article on Logical Increments, at the time of this research the FPS required for a smooth gaming experience is between 45-60 FPS (Logcial Increments, n.d.).

Therefore, for each AAI Technique, the simulated scenario will have its performance quantified using the Unity3D Profiling Tool and recorded for the duration of the simulation setting. Therefore, high performance is a quantified criterion.

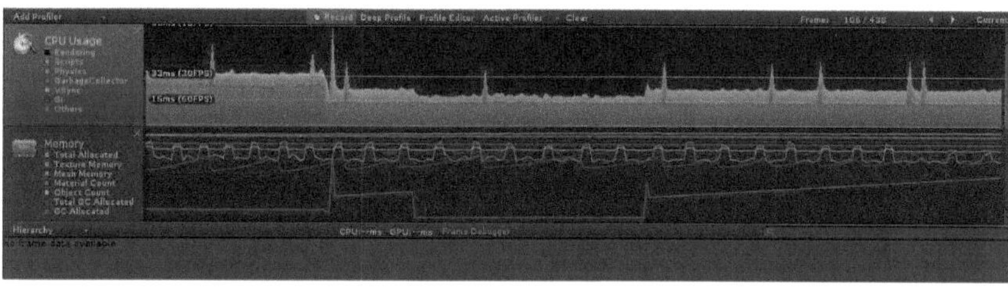

Figure 1 – the Unity3D Profiling Tool (Unity Technologies Inc., 2016).

Furthermore, the second criterion is **scalability**. Each AAI Technique will include a minimum number of agents in its simulation. This is due to the fact that background ambience often requires a significant of agents present in order to add depth to the user's experience (Graham, 2015). Scalability shall be measured by incrementally increasing the number of simulated agents without the performance decreasing below the benchmark. Lastly, the third criterion is **run-time adaptability**. each technique may require a set of its elements to be defined on edit time opposite to run-time in order to be executed. For example, in the AAI Technique "smart zones" it is required to define the roles for each smart zone before the living scene can start. Moreover, this criterion is considered qualitative as it is based on comparing the amount of elements that require to be defined at edit time for each technique, in order to qualify the most adaptable technique on run time

Simulation Scenarios: Design Overview

In this section, hypothetical scenarios are defined for each AAI Technique that resemble environments in RPGS.

AALS

To begin with, we define the scenario for AALS. The scenario for AALS shall be composed of 5 NPCS exhibiting a common ambient scene found in RPGS, the hypothetical scenario is that of a medieval village, each NPC will be portraying a simple ambient behavior such as speaking to other NPCS. this scenario is designed based on preference and speculation.

For this scenario, the ambient scene to be portrayed shall be composed of 5 NPCS each exhibiting a behavior. Three NPCs shall be exhibiting a conversation behavior with one another, a scene commonly found in most games. Secondly, an NPC shall exhibit a working behavior where he cleaning the front porch of his cabin. And lastly, a NPC shall exhibit a begging behavior in order to increase the diversity of ambience in the scenario.

Smart Zones

Secondly, for smart zones, a living scene is defined that exhibits a similar ambience to the scene defined for AALS. To begin with, each smart zone shall be composed of 2 roles, 1 Speaker and 1 listener. Furthermore, the NPCS shall exhibit an ambient scene of two villagers exchanging words outside of a medieval house.

Scheduling

Thirdly, for scheduling a template schedule is defined to resemble a neutral NPC in an RPG setting. As an illustration, the schedule of a farmer NPC is defined. Below is an example table indicating the sequence of behaviors in the schedule. The time values are merely an illustration.

Schedule Entry	Time
Leave House	3:00
Get Farming Equipment	3:15
Go to Farm	4:00
Begin Farming	4:10
Drop Off Farming Equipment	8:10
Go to Sleep	9:00

Table 1 - example NPC schedule

In conclusion, we've designed three hypothetical illustrations that can demonstrate the core concepts of each of the selected AAI Techniques. These illustrations shall be implemented and conclusive data shall be quantified from the Unity3D Engine during the simulation of each AAI Technique in order to compare each of the AAI techniques and determine their effectiveness.

Test Cases: Overview & Definition

In this section, tests are defined in order to gather data from the simulations for comparison. Furthermore, the metrics for the tests are defined and illustrated.

Test Metrics & Definitions

To begin with, we define the baseline metrics which shall be employed in each test. However, certain tests may introduce extra metrics. Below are baseline metrics:

Metric Name	Value	Unit
Performance Benchmark	45-60	FPS
Minimum number of NPCS	5	NPCS
Maximum number of NPCS	20	NPCS
Minimum simulation duration	10	Seconds
Maximum simulation duration	3	Minutes
Number of NPCS in session	5-20	NPCS

Table 2 - baseline testing metrics

In Addition, tests shall be conducted in sessions where each session is a collection of data from the metrics employed. Moreover, multiple sessions may be conducted depending on the type of test being conducted. Furthermore, in correlation with the above mentioned metrics the hardware specifications used for the test sessions are recorded first hand before any test sessions in order to validate the measurements based on the capability of the hardware. Below is a sample table of hardware specifications.

Component Type	Spec employed in the test machine
RAM	12 GB – DDR2
Graphics Card	GTX – 960M 2 GB
Processor	Intel core i5 (3rd Generation)

Table 3 - example hardware specs

General Testing Limitations

In an ideal scenario each simulation would progress for longer periods of time in order to gain thorough insight on the capabilities of each of the selected AAI Techniques. However, for the purposes of this research an illustration is satisfactory to provide an insight on to the potential challenges and advantages of each technique.

Performance Testing

Firstly, for each simulation the performance will be recorded for its duration. Furthermore, for each test session the following metrics are employed along with the baseline metrics:

Metric Name	Value	Unit
Average ms consumed per frame by the AAI technique	Recorded	(ms)
Number of test sessions	Recorded	Min/Sec

Table 4 - performance test metrics

Moreover, the overall performance of each illustration shall be plotted in a graph in order to compare the performance of each AAI Technique. Below is an example.

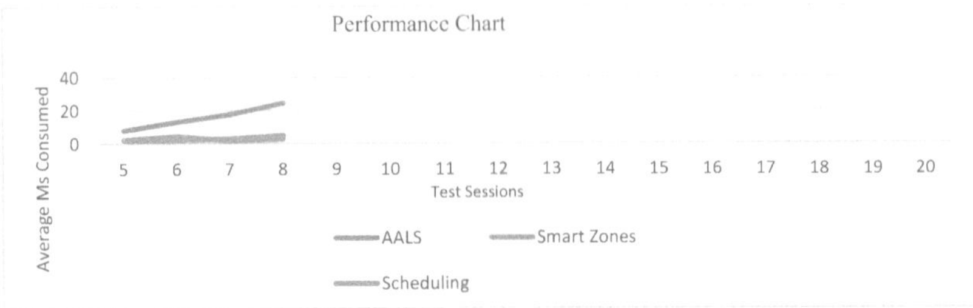

Figure 2 - example performance chart (Microsoft Office, 2016)

Scalability Testing

Moreover, as previously defined the "scale-ability" of each technique is the incremental increase in the number of NPCS per simulation without crossing the performance benchmark. In the context of research, "scale-ability" is the rate of performance deterioration relative to the number of NPCS being simulated during the session.

Furthermore, scalability shall be measured along with performance as the metrics for scale-ability are derived from the metrics used for quantifying the performance of each AAI Technique. The following metrics are employed for measurements:

Metric Name	Value	Unit
Difference in ms from previous session	Recorded	Ms
Number of NPCS in test session	Recorded	NPCS

Table 5 - scale-ability test metrics

Furthermore, the difference in ms is calculated per session. As an illustration, if a session was conducted with 5 agents then the agent count is scaled to 10 the difference is then calculated and plotted on the scalability graph. Below is an example graph.

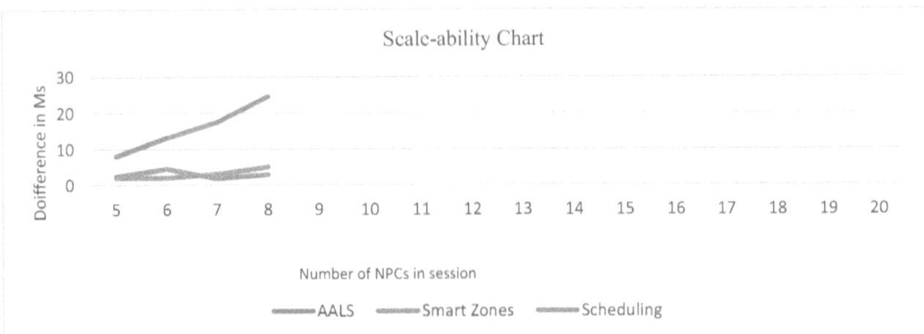

Figure 3 - example scale-ability chart (Microsoft Office, 2016)

Run-time adaptability Testing

Lastly, the final criteria for measuring the effectiveness of AAI Techniques is run-time adaptability, as previously defined for each AAI Technique a qualitative argument shall be constructed in order assess its adaptability in a generated context, similar to full-fledged procedural environments. Furthermore, each argument shall be utilized in a conclusive comparison in order to qualify the adaptability of each technique. Moreover, each argument shall be constructed based on the following key points:

- **Run-time independence:** the amount of edit-time configuration required.
- **Run-time variety:** the amount of changes required to create variance in the resulting ambience
- **Structural rigidity:** the amount of changes required to add new features.
- **Run-time content re-usability:** can the behavior be re-used by other NPCS or does each NPC require its own behaviors? As an illustration, in smart zones each smart zone defines the behaviors to be orchestrated but it does not define how many NPCS can be part of the living scene. Therefore, it is possible to re-use a smart zone for other NPCS that fit the role on run-time.

In conclusion, each test type shall be employed during the simulation period for each AAI Technique. Furthermore, the resulting mixture of quantitative and qualitative data shall be compared in a through comparative analysis that would conclusively illustrate the strengths and weaknesses of each illustration. Therefore, giving an insight into each approach employed by the selected AAI Techniques.

Simulation Scenarios: Implementation Overview

In this section, the simulations for the AAI Techniques demonstrated.

AALS

the designed scenario for AALS shall resemble that of a medieval village, a setting commonly found within RPGS. Below is an image of the virtual environment.

Figure 4 - AALS Environment (Unity Technologies Inc., 2016)

The Scenario

Each village shall be composed of 5 NPCS in accordance to the testing constraints. Each NPC shall exhibit an ambient animation loop in order to achieve collective ambience during the scenario. The ambient behaviors that were selected to be demonstrated are the following:

1) Speaking
2) Cleaning
3) Begging

Each ambient behavior was selected based on preference. Below are screenshots demonstrating each behavior in order to illustrate the desired collective ambience.

Figure 5 - AALS Speaking Behavior

(Unity Technologies Inc., 2016)

Figure 6 - AALS Cleaning Behavior

(Unity Technologies Inc., 2016)

Figure 7 - AALS Begging Behavior

(Unity Technologies Inc., 2016)

Illustration Implementation

Moreover, AALS shall be illustrated using the Animation System within the Unity3d Engine, each NPC shall be looping an ambient animation, below is a screenshot of the animation system in action.

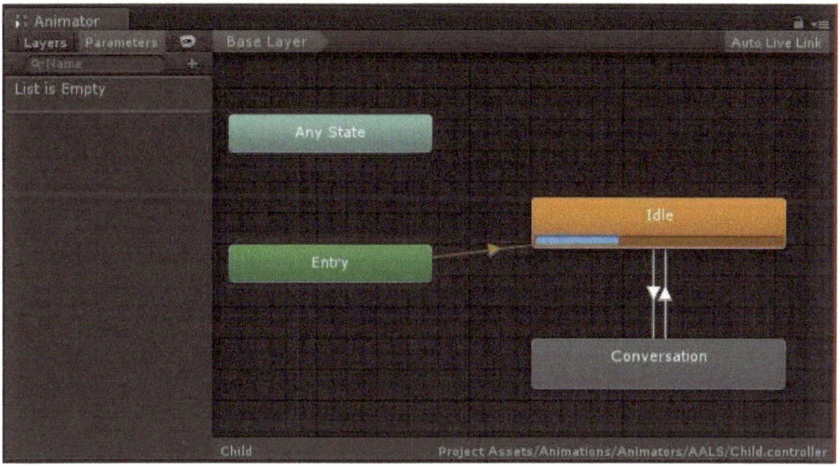

Figure 8 – an example Animation Controller (Unity Technologies Inc., 2016)

In the above screenshot, each block indicates an animation. Furthermore, the arrows indicate a transition between animations that can be based on a parameter or simply invoked once the previous animation is completed.

CCS: Overview

In Continuation, on run-time each NPC is spawned utilizing CCS. In this example, CCS would dictate that NPCS that are displaying a co-operative ambient behavior such as the animation "Conversation" in the above screenshot are spawned facing one another. In Addition, CCS dictates where to spawn NPCS who are displaying ambient behaviors that are singular such as the cleaning behavior previously demonstrated. Accordingly, CCS is implemented with the assistance of "spawn-points" placed during edit-time that indicate what type of NPC can be spawned. Therefore, based on the "spawn-points" CCS spawns any type of NPC with no limitations on numbers or types. This is for utilizing RTCG in order to produce a semi-PE. Below is a screenshot of the full AALS scenario.

Figure 9 - AALS Full Scene (Unity Technologies Inc., 2016)

Limitations & Modifications

For AALS, there was none encountered during development besides the lack of animation variety readily available for free usage on the Unity Asset Store. Furthermore, no modifications were required to be employed.

Scheduling

Secondly, the scheduling technique as previously described is a set of actions assigned to NPCS to be executed based on time. Each action is associated with a time frame for execution. Combined they are labeled as an "entry" in the NPC's schedule. Once the time frame of an entry expires, the schedule fetches an "entry" based on the current time, this continues until the schedule is fully executed (Graham, Breathing Life Into Your Background Characters, 2015).

The Scenario

For the purposes of illustrating the AAI Technique. A schedule was designed to resemble a glimpse of the daily routine of a medieval farmer. the virtual environment is a simple farm where each farmer executes the following schedule:

Entry Name	Time
Exit House and go to farm	0:00
Farm	0:15
Finish Farming and Enter House	Random

Table 6 - farmer NPC schedule

Furthermore, below is a screenshot of the virtual farming environment.

Figure 10 - Scheduling Farm (Unity Technologies Inc., 2016)

Moreover, below are screenshots demonstrating the farmer NPC executing each of the mentioned scheduling entries.

Figure 11- Farmer Walking Schedule Entry (Unity Technologies Inc., 2016)

Figure 12 - Farmer Farming Schedule Entry (Unity Technologies Inc., 2016)

Figure 13 - Farmer Re-Enter House Schedule Entry (Unity Technologies Inc., 2016)

Illustration Implementation

Furthermore, each Farmer NPC is spawned on run-time utilizing RTCG, each farmer is assigned his schedule upon spawning and depending on the current time the appropriate entry is fetched and executed. Below are the pseudo code steps for spawning NPCS.

```
Fetch npc spawn points.

Foreach spawnPoint in fetchedSpawnPoints

        Spawn: FarmerNPC at spawnPoint

        FarmerNPC: currentSchedule = farmerSchedule

End
```

Moreover, each farmer's schedule is executed per frame in order to constantly update the actions to be executed by the farmer NPC based on the current time. Below is a snippet of pseudo code highlighting the main steps in schedule execution.

```
If currentTime < currentScheduleEntryEndTime

        Execute [currentScheduleEntry]

Else if currentTime > currentScheduleEntryEndTime

        currentScheduleEntry = FetchNewEntryFrom [currentSchedule[currentTime]]

End
```

Limitations & Modifications

For the purpose of illustrating scheduling as an AAI Technique, a number of modifications were employed to the original implementation during development time.

To begin with, the implementation of the scheduling illustration follows the architecture provided in the original prototype developed by David "Rez" Graham. The scheduling technique was implemented as a data-driven prototype in C++ (Graham, Game AI Pro: Ch.36 Source Code, 2013).

this is due to the fact that the original prototype was purposely designed to be utilized by game designers. Furthermore, the original prototype introduced a number of different types of schedules and schedule entries. Providing a variety of choices for the implementation of different types of ambient behaviors. However, in the illustration the approach that was chosen is code driven as only the core aspects of the scheduling technique were implemented and it was developed in C# using the Unity3D Engine. Moreover, only the simplest form of schedules and schedule entries provided in the original prototype were implemented in the illustration.

This approach is due to the fact that the purpose is to provide insight into the approach employed by scheduling for simulating background ambience and not to develop a full-fledged implementation ready for commercial usage. Additionally, the development approach has no impact on the effectiveness of the AAI Technique as it is merely an implementation choice that does not change the behavior of the AAI Technique on an algorithmic level as well as not modifying its execution flow. Therefore, the AAI technique's approach and behavior are successfully replicated which as previously mentioned is the goal of the illustration.

Lastly, the illustration was developed with the architecture of the Unity3D Engine scripting system in mind, this introduced a potential disadvantage in performance for the following case. Each NPC is required to update their schedule entries based on the current time. For the sake of utilizing the Unity3D scripting system, this segment of the scheduling operation was completed in a standard update loop. However, this incurred redundant checking for updating entries as opposed to an event based approach which could have been more efficient but possibly ignoring the scripting system's architecture.

Smart Zones

Thirdly, the smart zones approach as it is described is an AAI Technique that orchestrates ambient behaviors on a scene level rather than an individual level such as the approaches found in AALS or Scheduling. Furthermore, each smart zone is assigned a list of required roles that will be portrayed by potential NPCS that enter the smart zone, if an NPC can fulfill a particular role, he/she participates in the living scene and is assigned a set of behaviors necessary to the living scene being orchestrated by the smart zone.

The behaviors associated with each role are orchestrated based on the duration of the living scene in order to produce a pre-defined sequence of actions that collectively demonstrate a coherent ambience (Cloderic Mars, 2015).

The Scenario

Furthermore, for the smart zone's illustration an ambient scenario is implement portraying two NPCS navigating to a medieval house then developing a conversation in front of its doors. Below is a screen shot of the smart zones environment. NPCs navigate from the shed in the screenshot before entering the smart zone surrounding the house.

Figure 14 - Smart zone environment (Unity Technologies Inc., 2016)

Moreover, below are screenshots demonstrating the execution of the living scene in the screenshot.

Figure 15 - agents navigating to nearest smart zone (Unity Technologies Inc., 2016)

Figure 16 - Agents occupying scene spots (Unity Technologies Inc., 2016)

And lastly, depending on the set time for the behavior orchestration to commence the living scene is executed, below is screenshot of the living scene in action.

Figure 17 - behavior orchestration in living scene (Unity Technologies Inc., 2016)

Illustration Implementation

The architecture of smart zones developed by Cloderic Mars is composed of 3 modules to handle the 3 major components of a smart zone: Triggers, Roles and behaviors (Cloderic Mars, 2015). The trigger module is responsible to validating the execution of a living scene, it checks if all the conditions required to execute a particular living scene are met such as the casting of all the primary roles that participate in the living scene. Secondly, the role assignment module is responsible for casting roles among the NPCS currently in the smart zone. Lastly, the behavior orchestration module assigns each NPC their respective behaviors based on the roles assigned to them, it then proceeds to orchestrate the behaviors based on the duration and time assigned for each behavior to be executed (Cloderic Mars, 2015). Below is a figure illustrating the architecture.

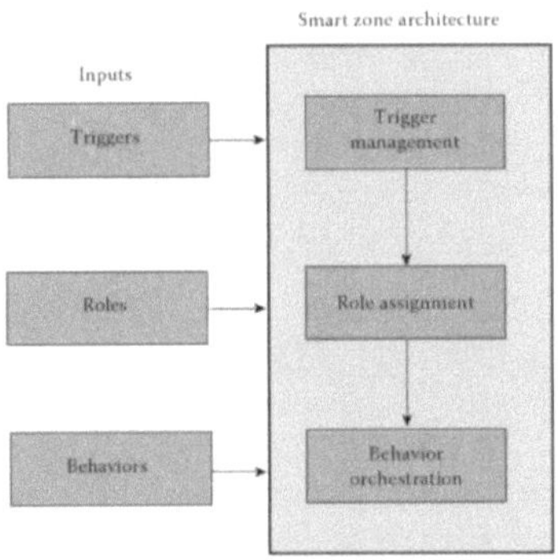

Figure 18 - Smart Zone Architecture (Cloderic Mars, 2015)

Execution Flow

Furthermore, based on the architecture described for the smart zone illustration a smart zone was developed with the following characteristics. For triggers the smart zone utilized a single trigger that would check if all the NPCS in the smart zone can fulfill the primary roles required to commence the living scene. Once the trigger is valid, the role assignment module steps in to cast the following roles:

1) Speaker: the NPC commencing the conversation.

2) Listener: the NPC listening to the other NPC.

Moreover, after the roles are correctly casted the behavior orchestration module is enabled. It first begins by assigning a behavior tree for each NPC that contains the behaviors ascribed to the NPC's assigned role. This behavior tree is dynamically loaded on run-time which gives a great deal of variety in behaviors as each smart zone can dynamically inject its own behaviors. This is in contrast to edit time development of the behaviors which would require each NPC to be assigned a behavior tree that includes all the behaviors of all smart zones, which may not be an ideal approach.

Limitations & Modifications

The smart zone illustration follows the original architecture developed by Cloderic Mars. However, similar to the other illustrations it only includes the core elements needed to demonstrate the smart zone approach to simulating background ambience. Therefore, the following modifications were applied.

To begin with, according to Cloderic Mars, the role assignment module's primary task is to cast the main roles required to execute the living scene then it would continue to cast other types of roles: Support and Extra. Both roles are not required to execute the living scene, however they are there to add to the ambience. This feature of the role assignment module was pruned from the illustration due to the fact that it is a secondary feature and not an essential one. It could possibly increase the visual fidelity of the smart zone illustration with the inclusion of more agents. However, the core execution flow of smart zones remains the same without its inclusion as the illustration still demonstrates the smart zones approach in a simplified manner.

Furthermore, similarly to the scheduling approach the original implementation of smart zones was developed with a data-driven approach. However, in the case of smart zones it is in the form of an editor built in the Unity3D Engine. This is due to the fact that the smart zones were developed as part of a larger AI Framework at MASA Life (Cloderic Mars, 2015) as well as the target audience being game designers. However, this would have been a valid feature to implement if the focus of this research was to developed an AAI System, but for the purposes of an illustration an editor is quite overkill.

Moreover, the original smart zones approach included a secondary system that simulated stats for NPCS. these stats were utilized in the context of allowing an NPC to decide if he would participate in a living scene or not. This is would be a beneficial feature to implement if the interest lied in investigating the compatibility of AAI Techniques with gameplay systems and how they can affect one another. However, this feature by nature required another feature to allow the NPCS to prioritize which smart zone to participate in if two were overlapping. Thus by extension this feature was pruned from the illustration as well.

Lastly, as for limitations a less than optimal approach for dynamically injected behavior in each NPC upon entering a smart zone was adopted due to misinformed assumptions. This resulted in a very limited form of injecting behaviors that did not allow for an array of methods to design each NPC's behavior tree according to the NPC's autonomous behavior as well as the NPC's orchestrated behavior.

Test Cases: Analysis of Data Gathered

In this section, data that is gathered from the test sessions conducted is evaluated in order to assess the effectiveness of the selected AAI Techniques. For each AAI Technique the following tests are conducted: Performance Test, Scalability Test & Run-time adaptability argument. Furthermore, the data is quantified from the Untiy3D Profiler.

To begin with, below is a table for the hardware specs utilized in the simulations.

Component Type	Spec employed in the test machine
RAM	24 GB – DDR3
Graphics Card	GTX – 980M 4 GB
Processor	Intel core i7 (3rd Generation)

Table 7 - hardware specs used for simulation

Furthermore, the following baseline metrics are employed in each test session. The metrics provide a rough perception of the potential resources used by each of the selected AAI Techniques

Metric Name	Value
Simulation Duration	30 seconds - 1 minute
Minimum Number of NPCS	5 NPCS
Maximum Number of NPCS	20 NPCS
Performance Benchmark	45-60 FPS
Number of Testing Sessions	4

Table 8 - test sessions baseline metrics used

Performance Testing

Each selected AAI Technique's performance is quantified in the following tables. In Addition, the quantified data shall be plotted on a final chart to compare the performance of each of the selected AAI Techniques.

AALS

For AALS the performance of Unity3D's animation system is quantified. Below is the table with the performance data for all the sessions.

Data Name	Session 1	Session 2	Session 3	Session 4
Average ms per frame	0.5 ms – 0.6 ms	0.9 ms – 1.1 ms	1.3 ms – 1.5 ms	1.7 ms – 2.1 ms
Simulation Duration	30 secs	30 secs	30 secs	30 secs
Number of NPCS in Session	5 NPCS	10 NPCS	15 NPCS	20 NPCS

Table 9 - AALS performance data

Scheduling

For scheduling, the performance of the scheduling code shall be quantified.

Data Name	Session 1	Session 2	Session 3	Session 4
Average ms per frame	0.16 ms – 0.30 ms	0.22 ms – 0.42 ms	0.49 ms – 0.58 ms	0.51 ms – 0.68 ms
Simulation Duration	30 secs	30 secs	30 secs	30 secs
Number of NPCS in Session	5 NPCS	10 NPCS	15 NPCS	20 NPCS

Table 10 - Scheduling performance data

Smart Zones

Similarly, for smart zones the performance of the smart zone code is quantified, it is composed of two elements: smart zone updates & agent behavior tree updates.

Data Name	Session 1	Session 2	Session 3	Session 4
Average ms per frame	0.24 ms – 0.39 ms	0.48 ms – 0.70 ms	0.65 – 0.82 ms	0.83 ms – 1.3 ms
Simulation Duration	30 secs	30 secs	30 secs	30 secs
Number of NPCS in Session	4 NPCS	10 NPCS	15 NPCS	20 NPCS

Table 11 - Smart Zones performance data

Lastly, below is the performance comparison of the selected AAI Techniques.

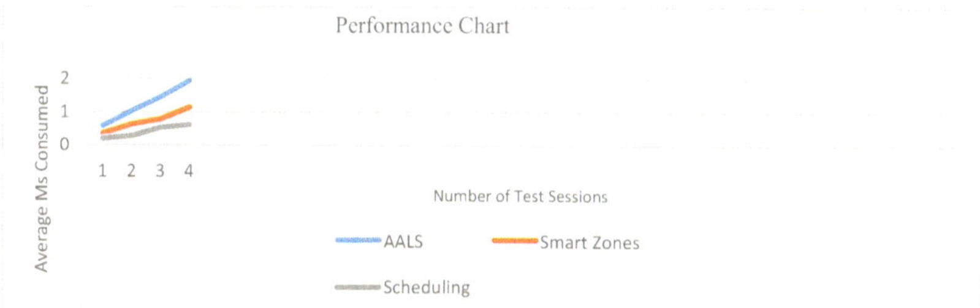

Figure 19 - Performance Comparison (Microsoft Office, 2016)

Scale-ability Testing

In continuation, each selected AAI Technique's scalability is measured and plotted in a scale-ability comparison graph. This graph should provide insight into the scale-ability of each technique. Furthermore, the following metrics are recorded for the test session.

Metric Name	Value	Unit
Difference in ms from previous session	Recorded	Ms
Number of NPCS in test session	Recorded	NPCS

Table 12 - Scale-ability testing metrics

The difference in ms is calculate as an approximate median in between the range of ms consumed within the testing session. For example, if the range is 0.22-0.30 ms 0.26 ms is the value where the difference is calculated from.

AALS

Data Name	Session 1	Session 2	Session 3	Session 4
Difference in ms from previous session	0 ms	0.4 ms	0.5 ms	0.6 ms
Number of NPCS in Session	5 NPCS	10 NPCS	15 NPCS	20 NPCS

Table 13 - AALS scale-ability data

Scheduling

Data Name	Session 1	Session 2	Session 3	Session 4
Difference in ms from previous session	0 ms	15 ms	20 ms	6 ms
Number of NPCS in Session	5 NPCS	10 NPCS	15 NPCS	20 NPCS

Table 14 - Scheduling scale-ability data

Smart Zones

Data Name	Session 1	Session 2	Session 3	Session 4
Difference in ms from previous session	0 ms	29 ms	14.5 ms	32.5 ms
Number of NPCS in Session	5 NPCS	10 NPCS	15 NPCS	20 NPCS

Table 15 - Smart Zones data

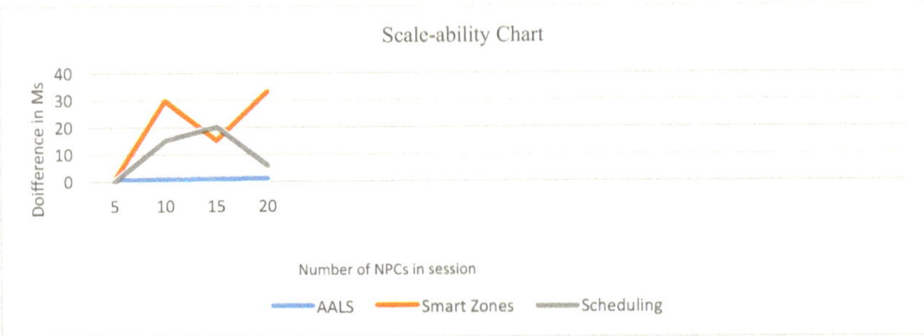

Figure 20 – Scale-ability Comparison Chart (Microsoft Office, 2016)

Run-time adaptability Testing

the final criteria for measuring the effectiveness of AAI Techniques is run-time adaptability. The qualitative argument constructed for each AAI Technique is mentioned below.

AALS

To begin with, AALS requires a significant amount of edit time configuration in comparison to the other selected techniques, this is mainly due to preparation of two components: CCS and animation loops. In fact, in absence of either of them AALS loses any form of run-time adaptability. Furthermore, AALS does not require a large amount of changes in order to gain variety, with the usage of CCS and switching the animation loops being displayed. AALS can provide a substantial amount of run-time variety. Moreover, since AALS does not require any implementation besides CCS. Therefore, the structural rigidity Is bound to the structure of the CCS spawner and the context where AALS is required to operate. Lastly, AALS does not provide any form of run-time content reusability as all the behaviors are defined on edit time and simply spawned in the correct context on run-time.

Scheduling

To begin with, scheduling only requires the configuration of the NPC schedules on edit time. Furthermore, a tremendous amount of features can be employed to create extremely diverse schedules and behaviors. According to David Rez Graham in his scheduling article, NPCS could have multi-action schedule entries which would increase the level of ambience involved. Additionally, Utility could be introduced to prioritize one action over another, releasing the rigidity of the schedules which would result in diverse variations among the NPCS (Graham, Breathing Life Into Your Background Characters, 2015). However, similar to AALS scheduling does not provide any form of run-time content reusability.

Smart Zones

To begin with, similar to the other AAI techniques, smart zones require initialization of their core components on edit time. Furthermore, to achieve variance in the resulting ambience, a number of options can be employed. To begin with, smart zones can be designed on diverse scales. One smart zone could only include 2 NPCS while another could orchestrate a much higher number. This level of control over the complexity of the overall ambience is not found in the other AAI Techniques. Furthermore, the structure of smart zones is quite rigid compared to the other techniques as it requires a linear execution flow, making slightly difficult to customize it. Moreover, smart zones are the only technique that can allow for run-time content reusability. This is achieved as NPCS who have completed a scene could simply navigate to another. Therefore, one could utilize a low number of smart zones with a higher number of NPCS being simulated.

Comparative Summary

In Summary, AALS consumed the most resources during performance testing. This metric is biased as each game engine's Animation system may consume a different load of resources. Furthermore, AALS was the only technique that scaled perfectly as there is no intensive operations being done during its simulation. Lastly, the argument for AALS provides insight that without the presence of CCS AALS would be a null choice for an effective AAI Technique in a PE. Furthermore, AALS provides no context for its behaviors as they are simply looped animations as well as providing a repetitive pattern for the user if any NPC is observed slightly. Thus potentially breaking the user's immersion.

Secondly, according to the data gathered during the simulations Scheduling was the most efficient in terms of resources consumed this is due to the time based nature of scheduling where only during scheduling entry updates are there resources to be used. Furthermore, scheduling was the 2nd best technique in terms of scale-ability. Furthermore, in comparison to the argument made for AALS each disadvantage of scheduling potentially had a solution. However, scheduling shares a similar problem to AALS in which it follows a repetitive nature through the schedules. If an NPC is observed the schedule pattern would be easily spotted but perhaps not as easily as AALS. However, in contrast to AALS the collective ambience produced by scheduling Is of higher quality as the variety of actions can be scaled according to the complexity of the AI required.

Thirdly, smart zones scored the worst on scalability as the values were quite unstable. Furthermore, it was 2nd best in performance. This is a natural result as smart zones are the heaviest of the AAI Techniques in terms of implementation involved as well as being the most complex of all three. Additionally, the argument for smart zones presented smart zones as the best of the three techniques in overall ambience as it operates on the agent level and the collective ambience level. Moreover, smart zones can result in potentially more diversity than the other techniques as they can be scaled to involve any number of NPCs without excess implementation. Accordingly, smart zones provide a level of reusability on run-time as their architecture does not require a specific NPC or a number of them rather it depends on the key abstraction of roles to allow for a high level of diversity not found with the other techniques as they mainly operate on the Individual NPC Level in contrast to smart zones.

Lastly, AALS is the simplest technique and scales tremendously, Scheduling can be considered an intermediary technique with balanced benefits and pitfalls. While smart zones are a great technique for controlling the ambience involved. However, it is quite unstable in scaling which may be posed as a problem in a commercial setting.

Research Conclusion

In conclusion, the goal of this research is to investigate contemporary AAI Techniques which can be adapted to suit a PG world. In light of the recent analysis of data gathered from testing each illustration, Each Technique portrays an array of benefits and weaknesses. Each illustration contains an insight on the potential development challenges that could be encountered when implementing any of these techniques in a commercial setting. Furthermore, the role of AAI in games has been defined through analysis of case studies and with each case study the role became clearer. As a result of this research 3 AAI Techniques were trialed and investigated in order to gain insight into the requirements of simulating background ambience in modern GAI. AAI is a tool for immersion and adding flavor to the experience of the user, when used appropriately it can exponentially increase the impact of interacting with NPCS on the user's experience. As a result of the comparisons conducted

In my Opinion, Scheduling is the median of all the techniques that were investigated, it is the most flexible due to its simple architecture and it was the more efficient in performance and scalability compared to smart zones which is the only other technique that was developed as a system. Furthermore, AALS is the cheapest AAI Technique however it provides very little immersion as it is quite repetitive in nature and requires CCS to be useful in a PE based on the investigation conducted for this research.

Finally, AAI is a young field in GAI. However, with the rise of techniques such as smart zones which seek to imbed complex behavior as a result of simulating ambience, the future of AAI may contain more realism and human like behavior to deepen the user's immersion further. Similarly, AAI is still in its early stages in PEs with little innovation or research conducted at the time of this research. However, similar to AAI in common environments. it is simply a matter of time before AAI exponentially grows into realism and become the most anticipated feature in a game after high fidelity graphics.

Future Research

In context of this research a few endeavors still remain in order to come closer to its applicability in the commercial world. To begin with, due to the investigative nature of this research, each AAI was only illustrated in order to portray its approach to simulating ambience. However, full-fledged implementations would result in more understanding of the capabilities of each AAI technique. Furthermore, utilizing a fully PE would provide relevant data that would contribute significantly to estimating the effectiveness of a AAI Technique. Moreover, higher amount of demo content would allow each AAI Technique to be exposed to different constraints. Therefore, thoroughly testing its scale-ability. Additionally, more complex simulations that potentially last for a significant amount of time would truly push each of the AAI Techniques to their limits providing more insight into each technique's true effectiveness in the face of real-world situations.

References

Adams, T. (2015). Simulation Principles From Dwarf Fortress. *Game AI Pro 2*, 549.

Buck, N. (2013). *Procedural Content Generation in Strategy/Role-Playing Games.* Minnesota: N/A.

Cloderic Mars, C. C. (2015). Smart Zones to Create the Ambience of Life. *Game AI Pro 2*, 550.

E.Goffman. (1959). *The Presentation of Self in Everyday Life.* New York.

Graham, D. R. (2013, September). *Game AI Pro: Ch.36 Source Code*. Retrieved from Game AI Pro: Collected Wisdom of AI Professionals: http://www.gameaipro.com/code/GameAIPro_Ch36.zip

Graham, D. R. (2015). Breathing Life Into Your Background Characters. *Game AI Pro 2*, 550.

Kevin Dill, B. L. (2014). *Game AI Pro.* Florida: CRC Press LLC.

Logcial Increments. (n.d.). *Information About Frame Rate*. Retrieved from Logical Increments: http://www.logicalincrements.com/articles/framerate

Microsoft Office. (2016, April 21). Microsoft Word.

Mike Lewis, K. D. (2015). *Game AI Pro 2* (2nd ed., Vol. II). (S. Rabin, Ed.) Florida, United States of America: CRC Press LLC. Retrieved 2016

Park, C. (2012, June 29). Enemy AI for Procedural Platformers in A VALLEY WITHOUT WIND. (AiGameDev.com, Interviewer)

PLch, T. (2014). *AI Engine behind Kingdom Come: Deliverance.* Retrieved from Nucl.ai: http://archives.nucl.ai/recording/ai-engine-behind-kingdom-come-deliverance/

Schwab, B. (2008). *AI Engine Programming.* Course Technology.

Unity Technologies Inc. (2016, April 21). Mechanim.

Unity Technologies Inc. (2016, April 21). Unity Scene Editor.

Unity Technologies Inc. (2016, April 10). Unity3D Profiler.

Varnier, J. (2014). *Far Cry's AI: A Manifesto for Systemic Gameplay*. Retrieved from Nucl.ai: http://archives.nucl.ai/recording/far-crys-ai-a-manifesto-for-systemic-gameplay/

Bibliography

Adams, T. (2015). Simulation Principles From Dwarf Fortress. *Game AI Pro 2*, 549.

Buck, N. (2013). *Procedural Content Generation in Strategy/Role-Playing Games.* Minnesota: N/A.

Cloderic Mars, C. C. (2015). Smart Zones to Create the Ambience of Life. *Game AI Pro 2*, 550.

E.Goffman. (1959). *The Presentation of Self in Everyday Life.* New York.

Graham, D. R. (2013, September). *Game AI Pro: Ch.36 Source Code*. Retrieved from Game AI Pro: Collected Wisdom of AI Professionals: http://www.gameaipro.com/code/GameAIPro_Ch36.zip

Graham, D. R. (2015). Breathing Life Into Your Background Characters. *Game AI Pro 2*, 550.

Kevin Dill, B. L. (2014). *Game AI Pro.* Florida: CRC Press LLC.

Logcial Increments. (n.d.). *Information About Frame Rate*. Retrieved from Logical Increments: http://www.logicalincrements.com/articles/framerate

Microsoft Office. (2016, April 21). Microsoft Word.

Mike Lewis, K. D. (2015). *Game AI Pro 2* (2nd ed., Vol. II). (S. Rabin, Ed.) Florida, United States of America: CRC Press LLC. Retrieved 2016

Park, C. (2012, June 29). Enemy AI for Procedural Platformers in A VALLEY WITHOUT WIND. (AiGameDev.com, Interviewer)

PLch, T. (2014). *AI Engine behind Kingdom Come: Deliverance*. Retrieved from Nucl.ai: http://archives.nucl.ai/recording/ai-engine-behind-kingdom-come-deliverance/

Schwab, B. (2008). *AI Engine Programming.* Course Technology.

Unity Technologies Inc. (2016, April 21). Mechanim.

Unity Technologies Inc. (2016, April 21). Unity Scene Editor.

Unity Technologies Inc. (2016, April 10). Unity3D Profiler.

Varnier, J. (2014). *Far Cry's AI: A Manifesto for Systemic Gameplay.* Retrieved from Nucl.ai: http://archives.nucl.ai/recording/far-crys-ai-a-manifesto-for-systemic-gameplay/